S. R Krom

Krom's Ore Crushing and Concentrating Machines

Plans for lixiviation and concentration works

S. R Krom

Krom's Ore Crushing and Concentrating Machines
Plans for lixiviation and concentration works

ISBN/EAN: 9783337164096

Cover: Foto ©berggeist007 / pixelio.de

More available books at **www.hansebooks.com**

KROM'S ORE CRUSHING AND CONCENTRATING MACHINES.

S. R. KROM, Engineer,

AND MANUFACTURER OF

ORE BREAKERS (Steel Sectional),

STEEL CRUSHING ROLLS,

PNEUMATIC CONCENTRATORS,

REVOLVING SCREENS,

GRAVITY DRY KILNS and ORE FEEDERS.

LABORATORY CRUSHERS,

LABORATORY ROLLS,

LABORATORY CONCENTRATORS,

LABORATORY SCREENS.

Plans for Lixiviation and Concentration Works,

Received the Highest Award from the U. S. Centennial
Commission for the Pneumatic Concentrator.

NEW YORK.

1885.

KROM'S IMPROVEMENTS

IN

ORE CRUSHING AND CONCENTRATING MACHINES,

Etc., Etc., Etc.   •

In connection with perfecting a system of pneumatic concen-centration, I had in view the improvement of machines for crushing and pulverizing ores. A study of the whole subject convinced me, that the principle upon which the construction of the Blake Crusher and that of the rollers are based, possessed the mechanical elements, which, if perfected and simplified, would make these machines the successful rivals of every other device. It is well known, that until very recently, the stamp battery held undisputed possession of the field, and mill men as well as mill builders, almost universally, believed that stamps would not be superseded by any other machine for fine crushing.

The first step to drive stamps from the field, was the introduction of rollers in the Bertrand and the Mt. Cory Mills. In both of these mills, the great superiority and economy of a system of crushing with rollers instead of stamps has been fully demonstrated.

While perfecting the rollers I had the fact confronting me, that they had frequently been tried without success. In fact, the universal opinion prevailed that rollers could be used only for coarse, and not for fine crushing. The secret of the difficulty, I discovered to be in the imperfect and weak character of the rollers heretofore used, due to a misconception of the requirements for a machine for such work, and a *failure to appreciate the advantages of continuous rolling crushing surfaces, operating on the toggle lever principle, which gives the greatest crushing pressure with the least power expended, and the minimum of wear.*

But no sooner is the fact established that rollers are destined to supersede stamps, than some persons jump to the conclusion that cheap rollers will also prove successful, and manufacturers of cheap machines are making statements that they sell rollers at one-half the price of my improved steel rollers, which will do the same work. Those who believe such statements and buy such cheap rollers, will pay dear for their experience.

It is the great strength of my rollers, superior material used, perfect workmanship, completeness and simplicity of design, which makes them a success and so economical; and no one understanding the subject will consider the price too high for the quality of the machines. As an aid to legitimate mining enterprise, it has been my aim to perfect a system of machinery for milling and concentrating ores, and to make such machinery a standard of excellence. The advantage and economy of employing the best which can be produced, is becoming slowly but better understood.

S. R. KROM.

NOTICE.

Some manufacturers are making ore breakers and crushing rollers which infringe on my patents. Purchasers of such machines are notified that they will be held equally liable for damages as the manufacturers.

All the machines, either ore breakers or crushing rollers which have tie bolts to take the strain due to crushing, are infringements on my patents.

S. R. KROM.

KROM'S
Ore Crushing and Concentrating Machines

ORE BREAKERS.

Every one who has any knowledge of ore crushing machinery is so familiar with the Blake Crusher and its construction, that no cuts or description of it is necessary here. But during the life of the patent (21 years) of this celebrated and valuable machine, the frame was cast in one piece, and no improvement of any importance was made in the machine until after the patent had expired.

Fig. 1.

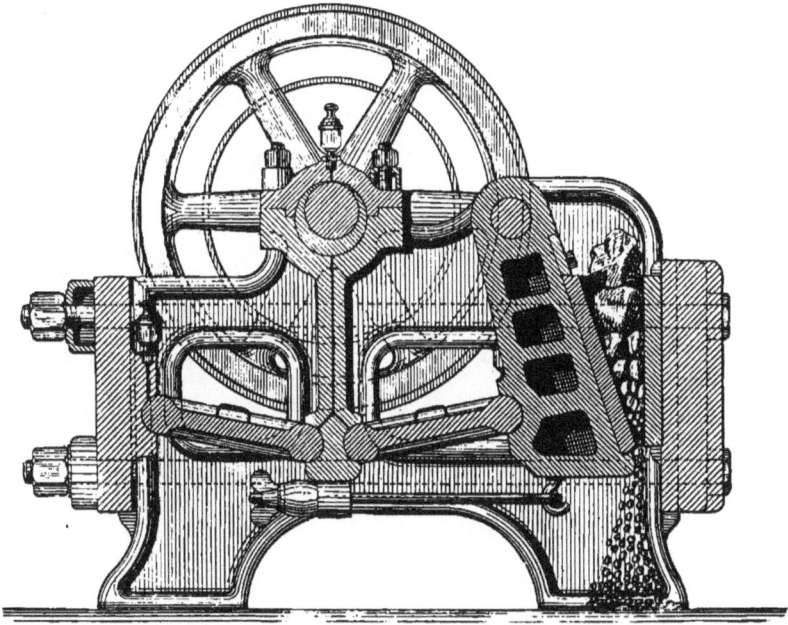

KROM'S ORIGINAL PATENTED SECTIONAL ORE BREAKER.
Re-issued April 11th, 1882.

The first practical plan to improve the construction of the Blake Crusher is shown in cut, Fig. 1, and was patented in 1875. The principal features covered by the patent granted for this machine (Fig. 1) are the tie bolts to take the strain due to crushing, and breaking-cups to relieve the machine from excessive strain. The breaking-cups are placed on the tie-bolts and also on the pitman, but it is intended that the small cups on the pitman should be the weakest and only these give away. A lug underneath the pitman holds it from dropping too low. This method of constructing a crusher with bolts not only makes a stronger machine, but *one of less weight,* and *easier* to transport, and allows the employment of means to prevent the machine from being subjected to undue strain.

Fig. 2.

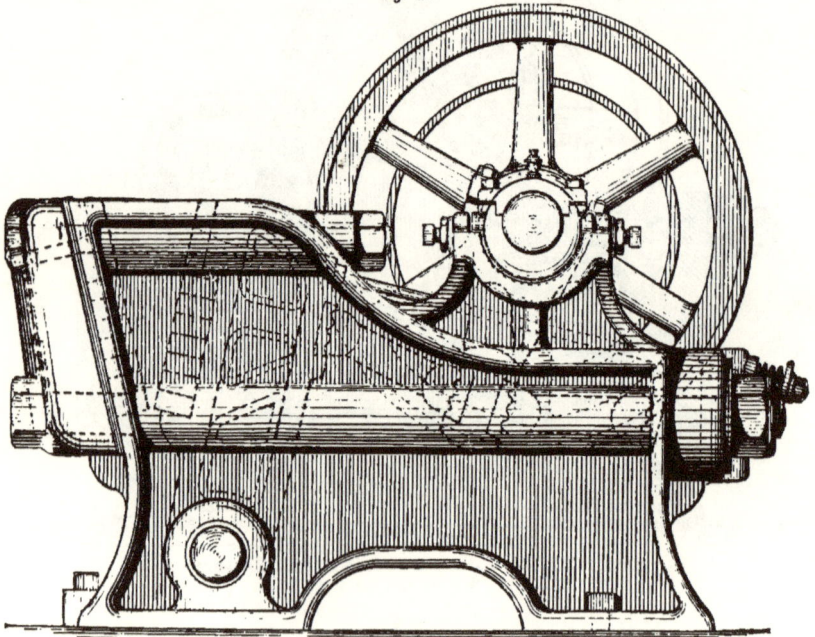

KROM'S IMPROVED PATENTED STEEL SECTIONAL ORE BREAKER.
Covered by Re-Issued Patent, April 11th, 1882, and Patent allowed for Improvements.

The next steps in improvements are shown in the cut,
Fig. 2. In this machine the upper tie-bolt is shortened
so as to give a better and more convenient form to the
side frame, and the lower tie-bolt is so placed that it
receives all the strain due to crushing the ore. In the
toggle abutment, through which the main tie-bolt
passes, are recesses around the bolt holes, and these
recesses are covered with wrought iron washers of
sufficient strength so they will not bend under the
ordinary strain in crushing the ore, but yield to ex-

Fig, 3.

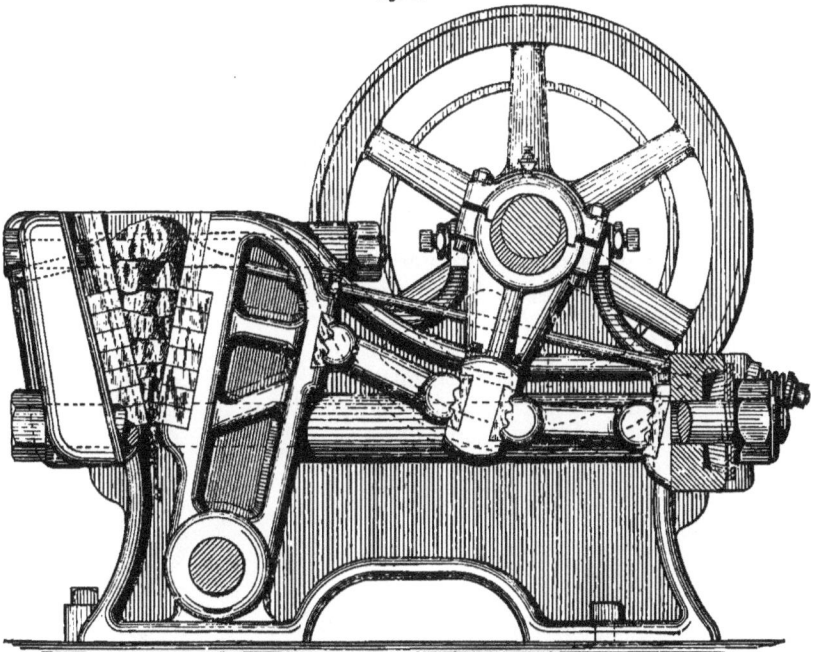

KROM'S IMPROVED PATENTED STEEL SECTIONAL ORE BREAKER.
Covered by Re-Issued Patent April 11, 1882, and Patent allowed for improvements.

8

cessive strain. These washers take the place of the
breaking cups in the first machine, but do not fly to
pieces, and therefore both ends of the toggle block yield
evenly together, and the frame of the machine is not
liable to be twisted or broken, or the bolts bent.

The third improvement is in forming the crushing
faces of bars of steel, and in the means of clamping
them securely in place. These bars are of good
steel, rolled to standard sizes, and cut to proper lengths.
The lower bars are hardened to increase their durability.
Provision is made for bringing the jaws closer together
to compensate for wear. Thin strips of metal are also
provided to put behind the bars to keep the wearing
faces in line. The bars or crushing faces can easily be
got at by taking the nuts from the bolts, and sliding
the stationary jaw forward. The fourth improvement
consists in hanging the jaws on an axis below the crushing
faces instead of at the top as in the Blake Crusher.
This manner of hanging the movable jaw gives a more
uniform product, and the principle is correct, as the
strain on the jaw is greater at the bottom than at the
top, and consequently the motion should be the least
where the strain is the greatest.

The fifth improvement consists in the employment of
toggles with rolling ends which work without friction or
oil. In this cut the teeth which hold the
toggles in place are omitted on one end so
as to better show the rolling surfaces.

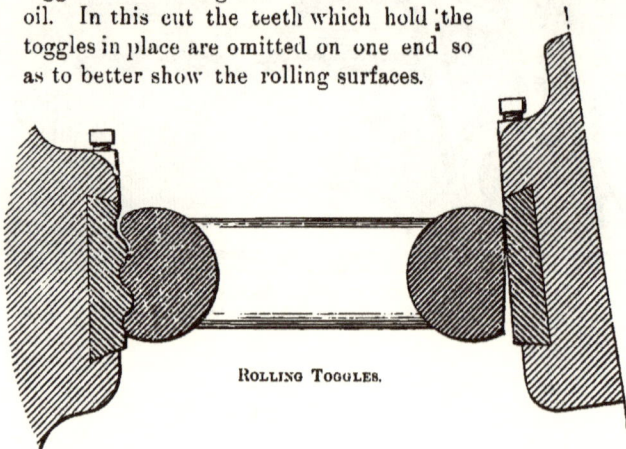

ROLLING TOGGLES.

The bearings are self-adjusting, large and long, and the machine is constructed for *high speed, hard work* and *large crushing capacity.*

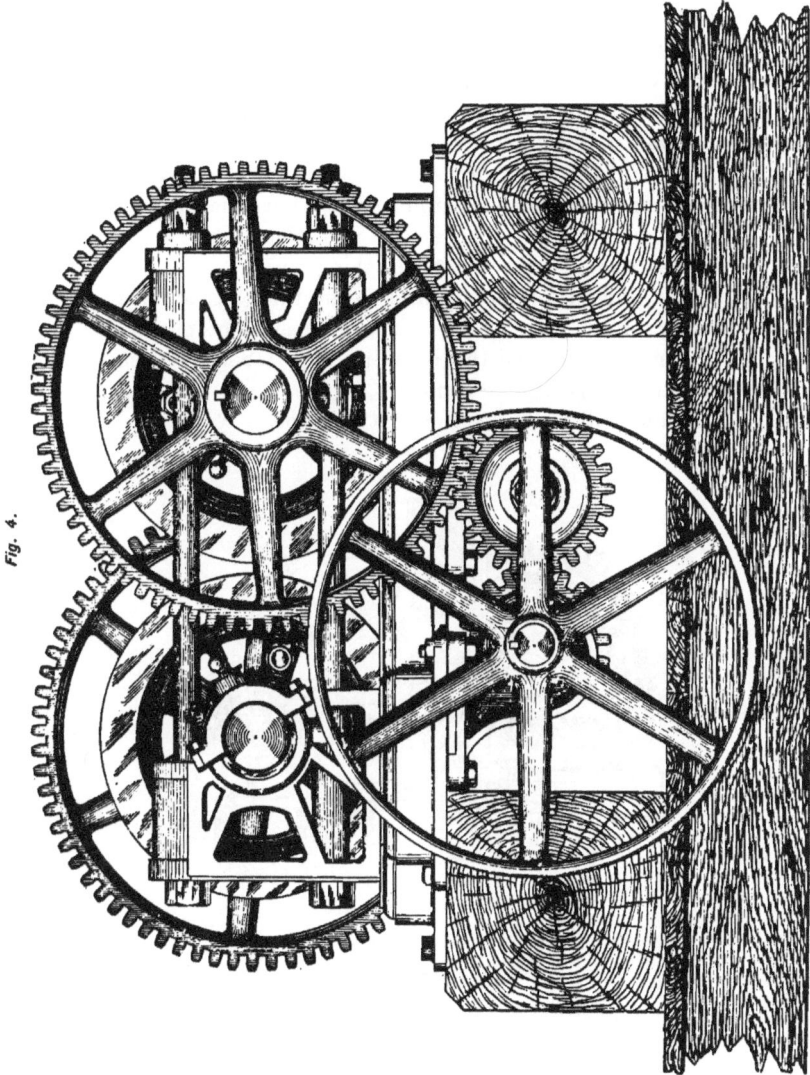

Fig. 4.

KNOX'S ORIGINAL STEEL CRUSHING ROLLERS.

CRUSHING ROLLERS.

Fig. 4, represents the first step in the improvement of crushing rollers constructed with bolts to take the strain. It is also the first crushing machine with forged steel tires. The system of gearing is also new, and an improvement on anything before or since it was introduced. It will be observed also that the pillow blocks are well adapted to take the strain. It is the first machine of the kind carried on a bed plate in one piece. Breaking cups are also introduced to relieve the machine from undue strain. This is the first time that such a device was introduced in crushing machines. The patent granted for these rollers covers the bolts to take the strain, the system of gearing, the manner of putting on the steel tires and the breaking cups. Some of the details in this cut are omitted as not being essential here, but will appear in the next machine.

Fig. 5.

.SIDE-VIEW

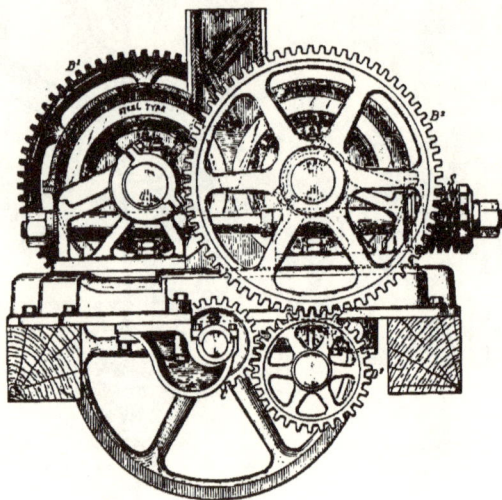

KROM's SECOND IMPROVED STEEL CRUSHING ROLLERS.
Covered by patent July 16th, 1872, and patents allowed for improvements.

The next step in the improvement of crushing rollers
is shown in cuts, Figs. 5 and 6. It consists of a recon-
struction of the pillow blocks so as to adapt them to
one bolt on each side instead of two, as in the previous
cut, Fig. 4. This at once simplifies the machine, and in
other respects makes a better pillow block. The bolts
being placed close to the bearings serve the purpose

Fig. 6.

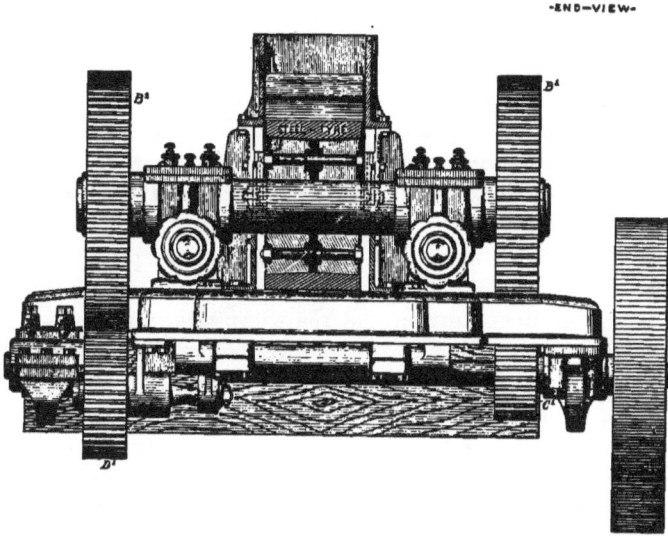

KROM'S SECOND IMPROVED STEEL CRUSHING ROLLERS.
Covered by patent July 16th, 1872, and patent allowed for improvements.

better than four bolts, two bolts being much more convenient for adjusting the rollers.

This arrangement also allows the shafts to be lifted
out of their bearings for repairs without removing any
of the tie-bolts, and permits the use of springs on the
tie-bolts, as shown at S, Fig. 5. It is not practicable to

use springs with four bolts. These springs were introduced to supersede the breaking cups in Fig. 4. The tie-bolts have solid collars near the middle, and a nut on each end. The springs and pillow blocks are placed between the collar and nut, so that the pressure is obtained without forcing the rollers together when the springs are set up to give the necessary resistance to crush the ore. At the date of this machine it was considered necessary to employ some means to relieve the machine from undue strain; and the springs as here shown are the best arrangement for the purpose yet devised. The arrangement of the gearing is the same as in Fig. 4, except that the intermediate gear D^1 is increased in size, and so placed as to require no change of position when the pillow blocks are moved up to adjust the distance of the rollers, and to compensate for wear. The wheel D^1 is placed nearly on a line drawn perpendicular through wheel B^2, but two inches in front of it; and when the tires are worn out the wheel D^1 will be two inches in the rear of the perpendicular line through wheel B^2. This movement of four inches, when so divided, only slightly affects the meshing of the teeth of the wheels D^1 and B^2, the variation being only $\frac{1}{32}$ of an inch from a true pitch. It will be seen that one roller is driven through the intermediate wheel D^1, and the other roller by a pinion, C^1, directly engaging with wheel B^1. By this arrangement and location of wheel D^1 the rollers can be adjusted without change of position of any of the gear. In the machine shown in Fig. 4, it was necessary to adjust gear-wheel D^1, as the gear-wheel B^2 was moved forward toward gear-wheel B^1.

It was of these rollers that Prof. Alex. Trippel spoke when he said—"The steel rollers of Mr. Krom's are exceedingly well constructed. They are undoubtedly the finest ever made, well proportioned, and very powerful. * * * An important fact which was developed on this trial, was the apparent practicability and advantage in substituting a system of crushing by steel rollers for stamp batteries."

(The above is from U. S. Commissioner Raymond's report of 1876.)

It was the success of these geared steel rollers at the Galena Mills, Nevada, that led to the adoption of the improved form at the Bertrand Mill at Geddes, Nevada, as described below.

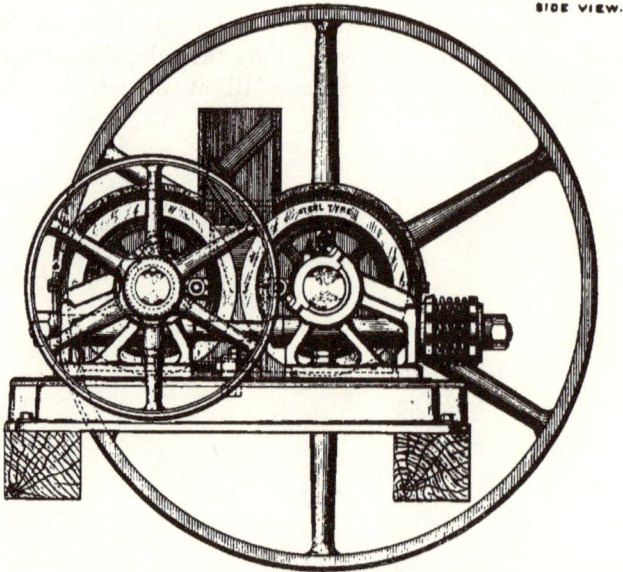

Fig. 7.

SIDE VIEW.

KROM'S THIRD IMPROVED STEEL CRUSHING ROLLERS.
Covered by patent July 16th, 1872, and patent allowed for improvements.

Figs. 7 and 8 illustrate very important additional improvements in crushing rollers. The 1st is the substitution of band wheels for toothed gearing. 2d. The substantial casing for inclosing the rollers, and a hopper to insure the spreading of the ore evenly across the face of the rolls. These improvements complete the adap-

tation of rollers for pulverizing ores to any degree of
fineness required. With pulleys we can run at any
speed desired, and thereby increase the capacity of the
rollers. The wear is reduced nearly to the crushing
faces alone. The danger of breaking the machine is
avoided, and the rollers perform their work with but
little noise and shock.

Fig. 8.

END VIEW

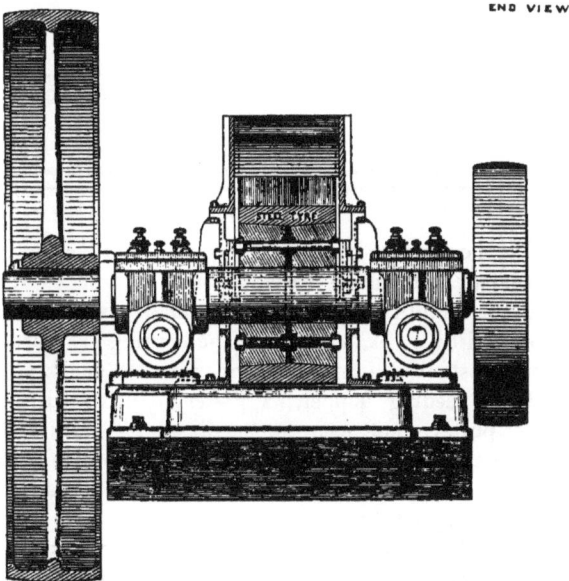

KROM'S THIRD IMPROVED STEEL CRUSHING ROLLERS.
Covered by patent July 16th, 1872, and patent allowed for improvements.

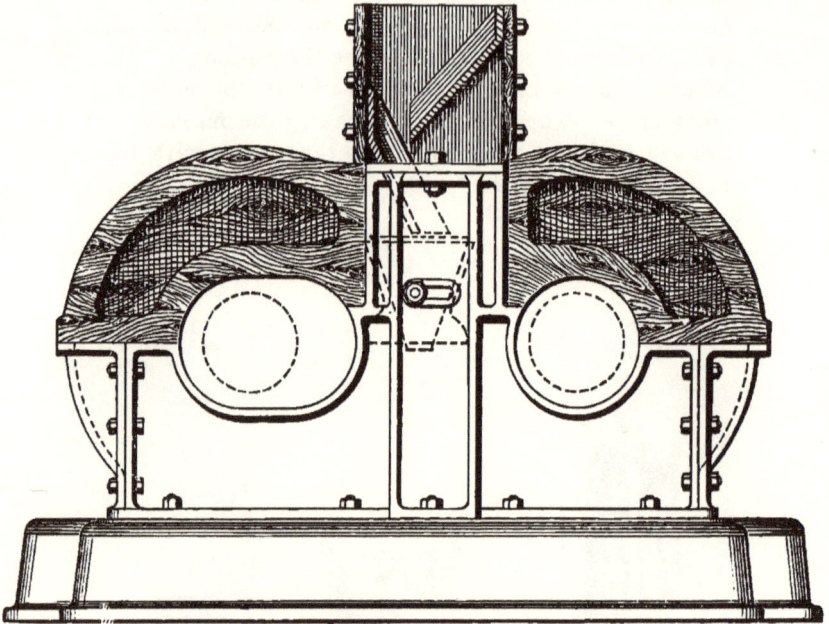

HOUSING FOR KROM'S PATENTED STEEL CRUSHING ROLLERS.

The housing (Fig. 9) so incloses the rollers that the
dust made in crushing is easily prevented from escaping
into the building by means of a small draught of air
produced by an exhaust fan. The housing also forms a
very substantial frame for supporting the cheek pieces
and the feeding hopper. I prefer to drive with one
large pulley, and use the small one to insure the
bite of the rollers upon the ore. Both rollers of course
when crushing travel at the same surface speed, but I
prefer to speed them so that the roller driven by the
small pulley, when the machine is not crushing, will
revolve one or two revolutions faster per minute. The

reasons for preferring this system of driving with only one large pulley, are :

1st. It would not be a good arrangement to hang on the movable pillow block, so large and heavy a pulley as would be required.

2d. The driving of one roller will cause the other to revolve at the same surface speed, where hard ore is fed between them, and therefore it is only necessary to drive one roller with a strong and steady power, and the other with sufficient force to insure the rollers always taking hold of the ore, and to keep the same in motion during the time when no ore is between them.

In reference to these rollers Mr. Stetefeldt said, before their success had been practically demonstrated:

"I am firmly convinced that Krom's Improved Steel Rollers, if they are given a fair trial, will make stamps a thing of the past. * * * * *

It is Mr. Krom's merit to have constructed rollers which reduce ores to any degree of fineness. The splendid results at the Geddes Mill, Nevada, have established the fact beyond any question. It is not too much to say that rollers produce the same results as the stamp batteries, with one-half the power and one-half the expense in wear, leaving their much lower original cost out of the question."

Mr. Albert Arents, Superintendent of the Mt. Cary Mill, says : "Now as to the economical question, the only one of practical interest, I will state that the cost of treating a ton of ore at the Bertrand Mill is thirty per cent. (30%) cheaper than at the Northern Bell Mill (a stamp-mill), although the latter is crowded to its utmost capacity, and treats more ore than any other mill of its size I know of. * * * * * I once advised having a side motion to one of the rollers, but I find if they are fed even the rollers wear even. I am astonished at the even degree of wear of the steel tires. I would not use stamps now for our purpose under any considerations."

Mr. R. D. Clark, Superintendent of the Bertrand Mining Co., says: "One great reason of our cheap milling at the Bertrand is that we use rollers instead of stamps. We crushed about 15,000 tons of ore before putting new tires on the finishing rollers, and have crushed in all about 20,000 tons, and we do not expect to put new tires on the roughing rollers for two or three months yet. From my experience the past fourteen (14) months, it is safe to calculate that each set of tires and each set of phosphor bronze linings will last to crush 20,000 tons of ore. There has been no other expense for repairs upon our rollers except the cheek pieces. The faces of the worn-out tires are not grooved at all, and they would be still good for wear had they not expanded by becoming worn thin. They have certainly stood the test of time, and are to-day more in favor with us than ever before. We would not have stamps if furnished without cost, and kept in repair for nothing also."

Additional testimony from an experienced mill man:

"I have been an advocate of stamps, but after seeing the quantity of ore your rollers have crushed at the Bertrand Mill (30,000 tons), I am convinced of their superiority over stamps, and have decided to use *them* in my new mill.

SIMEON WENBAN,

CORTEZ, NEVADA."

Much more could be added in the way of favorable comments in relation to steel rollers of the kind shown in the cut and just described in Figs. 7, 8 and 9, but further quotations would be unnecessary here, since the testimonials above are quite full, and come from persons widely and favorably known as the most skillful in the profession of mill-men and metallurgists.

Fig. 10.

SIDE VIEW OF KROM'S FOURTH IMPROVED STEEL CRUSHING ROLLERS.
Patented July 16th, 1872, and patent for improvements pending.

The next and latest step in the improvement of crushing-rolls is the swinging pillow-blocks, shown in Figs. 10 and 11. This, by reason of its uniqueness and importance, deserves to rank first in the order of merit. In all other machines one pair of the pillow-blocks are arranged to slide on the bed-plate, and each one of the sliding pillow-blocks has to be adjusted separately. It requires care to bring up the two movable pillow-blocks evenly and parallel with the stationary ones; and any looseness between the faces of the movable pillow-blocks and the bed-plate results in damage to the faces and pounding on the bed. In this machine all the bearings are securely and firmly fixed to the bed-plate, and no chance whatever exists for looseness, causing wear or damage to any part of the machine. (In fact, the movable bearings are now better secured to the bed-plate than the stationary ones, although the latter are very securely bolted on the bed.) The two swinging bearings are united together by a very heavy shaft (m, Fig. 11) 11 inches in diameter, so that the bearings must swing together, and consequently the rollers and shafts are always parallel. The shaft m, connecting the swinging bearings, is so strong that one bolt on one side is sufficient to hold the rolls together when crushing ore. It will be observed that springs have been dispensed with. The machine is constructed so strong that they are no longer required; and besides I now employ magnets to take out from the ore all pieces of steel which might dent the faces of the rollers.

The rolls work better and more economically without springs. Contrary to what might at first be expected, the machine is more simple than any other and more durable. The tires are held by two heads slightly cone-shaped. One of these heads is securely fixed to the shaft by shrinking it on. The other is split on one side,

Fig. 11.

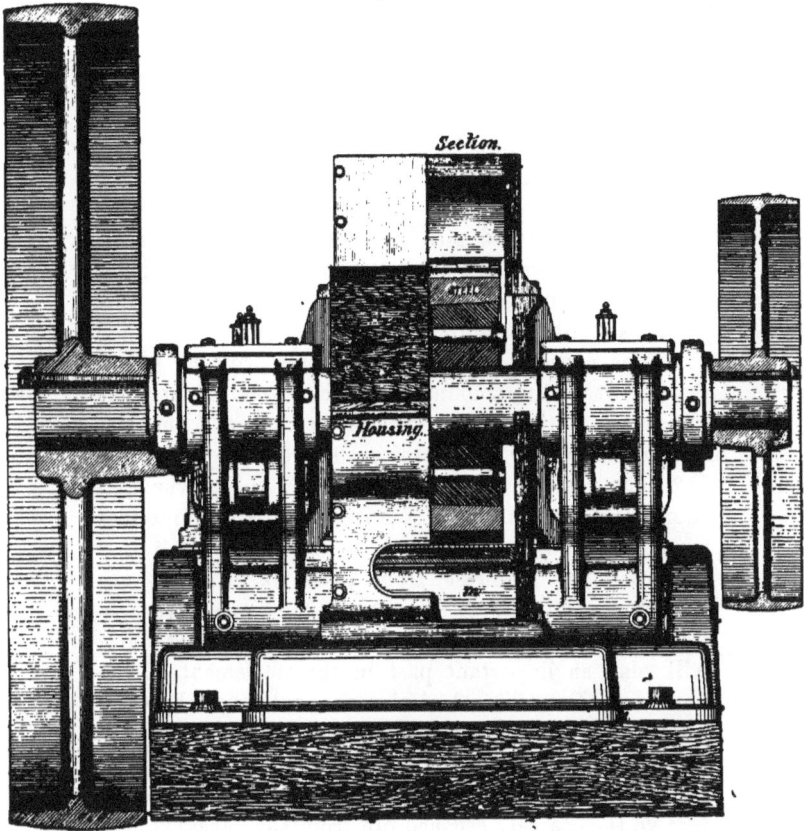

END VIEW OF KROM'S FOURTH IMPROVED STEEL CRUSHING ROLLERS.
Patented July 16th, 1872, and patent for improvements pending.

so that when the heads are drawn together by the bolts, within the tires, the split head will close tightly upon the shaft.

Recapitulating the several improvements described we have:

1. Steel tires and the method of securing them.
2. Pulley-gearing.
3. Housing to inclose the rollers.
4. Swinging pillow-blocks.
5. Bolts to take the crushing strain.
6. The hopper for spreading the ore.

This paper would not be complete without a comparison between rolls and stamps (other pulverizing machines not being worthy of attention here). On this head I submit the following observations:

1. The capacity of any machine for crushing ore is dependent on the amount of crushing surface brought in contact with the ore in a given time, and the efficiency with which the crushing surface is applied.

2. The cost of crushing depends on the power consumed and the expense of maintaining the machine in a working condition.

It will be manifest from these axioms that the employment of correct mechanical principles with simplicity of construction and fewness of working parts will play an important part in the settlement of the question. For this calculation we will begin by computing the capacity of two sets of 26-inch rolls with 15-inch face, the average diameter of which by reason of wear will be say 24 inches. One pair of the rolls is to receive the coarsely crushed ore from the breakers, and the other set is for finishing or pulverizing. Such rolls, running 100 revolutions per minute, will bring into contact with the ore 113,100 square inches, and the two sets of rollers are therefore equal to 226,200 square inches of effective crushing surface per minute. .

The usual diameter of each stamp-shoe and die is 8 inches, and the area is therefore 50 square inches to each stamp. *Counting all this as effective crushing surface*, we have for each stamp (falling say 90 drops per minute) 4,500 square inches of crushing surface. It will therefore require 50 stamps to give nearly the same surface (namely 225,000 square inches) as can be obtained with two sets of 26-inch rollers.

Let us now consider the two machines from a purely mechanical point of view. Stamps crush by the percussive effects of a falling weight, and the power consumed is in excess of the work performed, because it takes the same power to operate stamps, whether the blows are effective or not, and it is safe to assume that one-half the power is lost. With rollers the crushing force is applied continuously in one direction, and, other things being equal, it must be evident that the power is far more economically utilized. Fig. 12 illustrates the difference, representing the faces of the rolls as composed of the shoes of a stamp battery, and the stamp-heads, stems and tappets as fastened to or composing the fly-wheel, to give weight and steadiness of motion and crushing force sufficient to overcome irregularites of resistance. It must be evident that if the rolls (Fig 12) are held sufficiently rigid together, the pieces of rock falling between them will be crushed and the power consumed will be in proportion to the work done, while the surplus power will be stored up in the fly-wheel. If the same amount of metal were put in stamps on the other hand, and the stamps did not (as of course they would not) give in falling a full equivalent of useful work for the power consumed in raising them, the excess of power is wasted with no possibility of recovery. Moreover, the blow of a stamp has no regular efficiency, and is always only partially effective by reason of crushing on too much ore or on ore already fine enough.

With rolls the crushing effect is positive, and ore

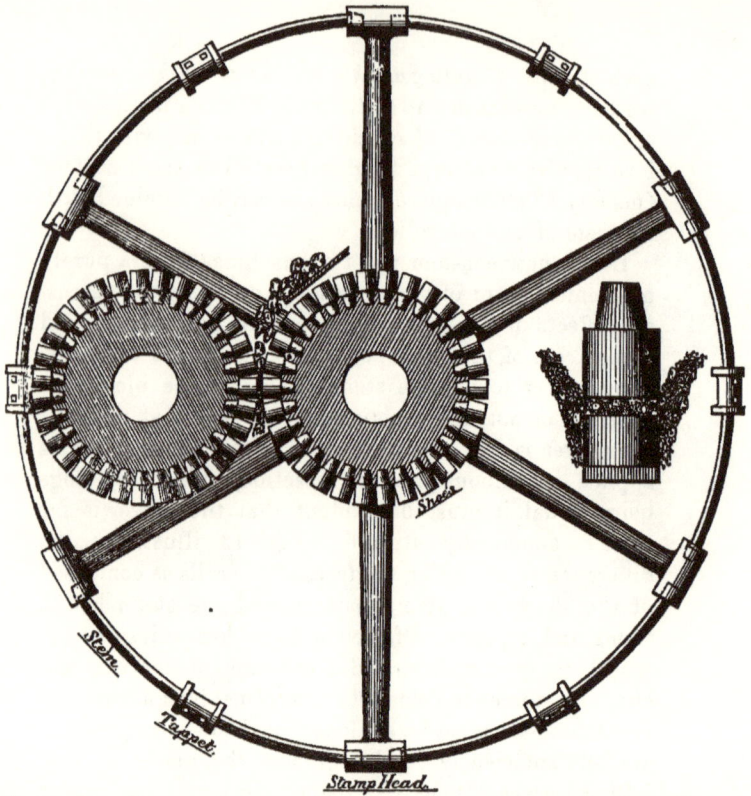

Fig. 12. IDEAL FIGURE TO ILLUSTRATE RELATIVE ECONOMY OF POWER IN STAMPS AND ROLLERS.

Fig. 13. THE TOGGLE-LEVER PRINCIPLE COMPARED WITH FALLING WEIGHTS.

cannot pass between them without being crushed if they are properly held together. Besides, no ore which is fine enough is acted upon a second time. Again, the toggle-lever principle acting in rolls, as illustrated in Fig. 13, serves to give the necessary crushing pressure with the smallest expenditure of power, while with stamps the contrary is the case.

I have met persons who have claimed that the percussive effect of a blow has some quality more effective than pressure. Now the velocity which the stamps attain at the moment of striking the ore, dropping 9 inches, is 7 feet per second. while the surface of rolls 24 inches in diameter, running 100 revolutions per minute, travels 628 feet per minute, or about 10½ feet per second. This gives already more percussive effect than stamps; so that, if percussion is any aid to pulverizing, we have it in the rolls also by giving them velocity. But I do not adopt this reasoning. The whole matter is summed up in the statement, that stamps *must* attain a certain velocity to crush, while the crushing effect of rolls is produced at either high or low velocity, without material change in quality of product; speed, however, giving increased capacity.

Simplicity of construction has also been claimed for the stamp-battery. Let us see if this claim can be sustained.

The working or wearing parts of each set of two rolls, as above illustrated, are as follows:

4 journals on two shafts.
4 pillow-blocks for journals.
2 crushing-rollers.
2 side wearing-plates.

Total: 12 wearing-parts to each set of rolls or 24 in two machines. None of these parts can break, or are subject to rapid wear.

On the other hand, the working or wearing parts in a 50-stamp battery are:

15 journals on 5 shafts (viz., 10 stamps on each shaft).
5 cam-shafts.
15 bearing-boxes for journals.
50 stems.
100 guide-boxes for stems.
50 stamp-heads (bosses).
50 shoes.
50 dies.
50 cams with keys.
50 tappets with keys.

Total: 435 parts, all of which are subject to severe wear or liable to break.

Recapitulating our comparison:

1. The average crushing surface of 26-inch rolls is 226,200 square inches, while the crushing surface of 50 stamps is only 225,000 square inches per minute.

2. Two sets of rolls have only 24 wearing parts, while a 50-stamp battery has 435 wearing parts (to say nothing about the screens which are frequently damaged, and the meshes of which become clogged with ore).

3. The breakage and wear of stamps are much greater than that of rolls.

4. The rotary crushing motion of rolls, mechanically considered, is superior and more economical than crushing under falling weight, or with any other machine or system.

5. By the toggle-lever principle in rolls, advantage is taken of the most economical and powerful means of obtaining the necessary crushing-pressure on the ore, which cannot be obtained in any other way, and is not obtained in any other machine.

Mr. C. A. Stetefeldt, in an appendix to his able paper on *Russell's Improved Process for the Lixiviation of Silver*

*Ores,** has made a comparison between steel rolls and stamps, which I here reproduce.

"COMPARISON BETWEEN KROM'S ROLLS AND THE STAMP-BATTERY.

The successful introduction of Krom's rolls in the Bertrand lixiviation mill, Nevada, is of such importance that it deserves notice here. In comparing stamps and rolls we have to consider : 1st. The physical difference of the pulp produced. 2d. The question of economy, namely : *a.* In regard to original cost of constructing the plant. *b.* The wear and tear of each machine, and consumption of fuel.

1. *The Physical Difference of the Pulp Produced.*

I refer here to the well known fact, that if pulp produced by rolls or by stamps is sifted through the same size of screen, the ore-particles from the former are more uniform in size, and contain much less impalpable dust than those from the latter. It has been found that for chloridizing-roasting great fineness of the ore is entirely unnecessary—provided the silver minerals are not too finely impregnated in the gangue. In the lixiviation process great fineness of the ore interferes with rapid filtration. From this it follows that ore pulverized by rolls is in the most favorable condition as far as the mechanical part of lixiviation is concerned.

2. *The Question of Economy.*

A discussion of this subject, which is complete and thorough, and compares the efficiency of rolls and stamps under varying conditions, is not possible at present, because the available statistics concerning Krom's rolls are confined to those from the Bertrand Mill. Sufficient evidence, however, has accumulated to prove the superiority of the rolls beyond any doubt. Their introduction in the Mt. Cory Mill, Nevada, will soon bring additional proof. It seems to me that the application of rolls is most favorable in cases where silver is extracted by lixiviation, and the character of the ore permits comparatively coarse crushing without interfering with good roasting.

A comparison between rolls and stamps will be made from the following premises: I assume that the crushing capacity of two

* *Transactions of the American Institute of Mining Engineers,* vol. xiii., p. 114.

sets of Krom's 26-inch rolls is equal to that of a 30-stamp battery with stamps of 850 pounds, dropping from 7 inches to 8 inches ninety-four times per minute. Mr. Clark, superintendent of the Bertrand Mill, states that he can crush, with two sets of rolls, 100 tons of ore in twenty-four hours to such a fineness that all will pass through a No. 16 wire screen, consuming not over 4 cords of wood for power. The ore has a quartz gangue, and is by no means an easy crushing ore. The fuel required for running 30 stamps is about 6 cords of wood in twenty-four hours. For some remote locality in the West the following prices are assumed, namely: Freight at 3 cents per pound; lumber at $50 per thousand feet; wood at $6 per cord; wages of carpenters at $4.50, and of mill-wrights at $6. Certain items of construction will be about equal, namely: conveyors, elevators, revolving screens, and dust-chambers. Revolving screens are also used in connection with a well-appointed battery in order to separate coarse material resulting from the breakage of battery-screens. The building, however, for rolls will be much smaller than that for the battery, and a saving of not less than $1,500 will be effected in its construction. Finally, the rolls requiring less power, a saving of at least $1,250 will be made in providing and setting up engine and boilers in a mill with rolls

Cost of Erecting a 30-stamp Battery.—The plant, including hard wood screen-frames and guides, wooden pulleys on cam-shafts, Tullock's feeders with iron hoppers, and all necessary bolts, weighs 90,600 pounds, and costs in Chicago $5,850, according to a statement of Messrs. Fraser and Chalmers. The framework takes about 36,000 feet of lumber, and the expense of setting up the battery is estimated at $4,000. Hence, total cost of constructing a 30-stamp battery is:

Plant at foundry,	.	$5,850 00
Freight, .	. .	2,718 00
Lumber,	. .	1,800 00
Cost of setting up,	.	4,000 00
		$14,368 00

To this has to be added in order to compare with rolls:

Extra cost of building,	. . . . .	1,500 00
Extra cost of engine and boilers,	. . .	1,250 00
Total,	. .	$17,118 00

Cost of Erecting two sets of Krom's 26-inch Rolls.—The amount of lumber required for setting up the rolls alone is merely nominal.

From this it follows that also the labor of placing the rolls must be trifling. The weight of one set of 26-inch rolls is 12,000 pounds, and the cost in New York is $2,250. There is only one self-feeder required, and its weight is estimated at 2,000 pounds, cost $200.

From these figures we deduce the following:

Plant at foundry, ...	. $4,700 00
Freight, ...	. 780 00
Cost of setting up, including lumber,	. 700 00
Total, . .	. $6,180 00
Difference in favor of rolls:	
	$10,938 00*

Wear and Tear of Stamps and Krom's Rolls.—In comparing the wear and tear of stamps and rolls we cannot very well express it per ton of ore crushed, because the capacity of the pulverizing machinery is a function of the hardness of the ore, and of the fineness of the pulp produced. A much correcter method will be to take the figures per running time of twenty-four hours. Making estimates from this standpoint it is supposed that the wear and tear in running the machinery at full capacity is a nearly constant quantity, while the capacity is variable as stated above. That this assumption is correct for practical purposes has become evident to me by a comparison of the battery statistics from the Manhattan and Ontario mills. The conditions in these mills are by no means the same. There is a difference in the character of the ore; the Manhattan uses No. 50, and the Ontario No. 30 screens; the Manhattan stamps weigh 1,000 pounds, those at the Ontario 850 pounds; Ontario stamps drop 92 times per minute, those at the Manhattan about 100 times; still, in comparing the wear and tear per actual horse-power expended, the figures are very nearly the same. Hence, my argument does not lack the support of practical experience. The wear of rolls is principally confined to the steel tires, that of the battery to a great number of parts. With rolls the steel tires can be consumed to within less than one-half inch of their thickness, while with stamps, the shoes and dies have to be exchanged after only two-thirds, or less, of their weight has been worn, leaving other parts out of consideration. Another point should not be overlooked. The complicated construction of the battery causes considerable expense in skilled labor for repairs, which, in the case of rolls, is merely nominal. Advocates of the

*In both estimates elevators, conveyors, and revolving screens are not included, as stated previously.

battery have argued that its great advantage is the continuance of its operation if one battery of five stamps gets out of order, while both sets, or three sets of rolls, as the case may be, have to be stopped if repairs are needed for one set. But it is just the solid construction of Krom's rolls which reduces stoppages from this cause to a minimum. The system of elevators, screens, hoppers and conveyors, if properly constructed, will get out of order very rarely. How often it is necessary to hang up stamps for repairs is too well known to require any statistical proof.

Wear and Tear of Krom's Rolls.—As to statistics of wear and tear of Krom's rolls I am confined, at present, to those from the Bertrand Mill. Mr. R. D. Clark states that two sets of steel tires crushed in round figures 20,000 tons of ore.

As stated previously, the full capacity of the rolls is, in twenty-four hours, 100 tons, the ore being sifted through a No. 16 screen. In the beginning, however, the ore was crushed much finer, namely, so as to pass a No. 20 screen, and the daily capacity of the rolls was much less. Taking this into consideration, the actual wearing capacity of the tires cannot be estimated at less than 250 working days. The cost of this wear is as follows:

Two sets of steel tires at New York,	$764 00
Freight on 3,264 pounds, at 3 cents,	98 00
Total, ...	$862 00
Wear and tear per twenty-four hours:	
In steel tires, ...	3 45
In other parts, screens, supplies and lubricants,	1 75*
Wages for repairs, ...	1 25*
Total, . . .	$6 45

Wear and Tear of Stamps.—I have been favored with correct statistics from three of the most prominent mills in the West, namely: the Manhattan, Nevada; the Ontario, Utah; the Lexington, Montana. Taking into consideration the somewhat abnormal conditions at the Manhattan Mill, in so far as the weight of stamps there is 1,000 pounds, and the number of drops is greater than in either of the other mills, causing a more frequent breakage of stems and cam-shafts; further, that the statistics from the Lexington Mill are those from the first year's run, where certain breakages are reduced to a minimum; finally, that freight in these

---

*These figures will, no doubt, be considered too high by Mr. Krom and Mr. Clark. I consider it safe, however, to provide a limit for accidents.

localities, on account of direct railroad communications, is slightly less than I have assumed in my premises, I arrive, by making such allowances, at the following figures for wear and tear of a 30-stamp battery per twenty-four hours running time:

In all parts subjected to wear and breakage, supplies, screens, and lubricants,*	$11 50
Wages for repairs,	5 50
Total,	$17 00
Wear and tear of rolls,	6 45
Difference in favor of rolls, .	$10 55

Interest and Amortisation.—In comparing the expense of running rolls and stamps, interest and amortisation on the excess of capital required in the original construction of the plant for stamps cannot be neglected. Considering the short life of most silver mines in this country, this item should not be taken at a lower rate than 15 per cent. per annum. If we take the running time of a mill at three hundred and fifty days in the year, and consider that a mill with stamps will cost $10,938 more than one with rolls, the interest and amortisation amount to $4.68 per day.

Summary.—From the above we find the following daily saving in a mill with two sets of Krom's 26-inch rolls as compared with 30 stamps:

Wear and tear, and repairs,	$10 55
Interest and amortisation,	4 68
Fuel, 2 cords of wood, at $6,	12 00
Total, .	$27 23

If no great accuracy can be claimed for this estimate, it is the best which can be given at present.

Mr. Krom, supported by Mr. Clark, claims that two sets of rolls will crush more ore than 30 stamps. Others will consider my estimate too much in favor of the rolls. Time will establish the correctness or fallacy of these views."

*Of this amount, the wear and tear of shoes and dies represents only 40 per cent.; tappets, bosses, cams, stems, cam-shafts, flanges, and boxes, 39 per cent.; and screens, lubricants, screen-frames, battery guides, and carpenters' and machinists' supplies, 2½ per cent.

To the comparison of Mr. Stetefeldt I need add nothing except to corroborate and emphasize it from later data of experience.

Since Mr. Stetefeldt's paper was written, the capacity of the rolls in the Bertrand mill has been rated at 150 tons per twenty-four hours crushing to pass a 16-mesh screen. This is 50 per cent. more than Mr. Stetefeldt felt justified in assuming. The rolls have also been put in successful operation in the Mt. Cory mill, Nevada, and the Haile mill, South Carolina.

What the full capacity of the improved rolls is, has not yet been demonstrated by actual test, as either the screening or elevator capacity has not been sufficient to allow a complete demonstration of this point. We only know that at the Bertrand mill the rolls crushed 50 tons of hard ore in twelve hours, and in the Mt. Cory mill, 50 tons to 30-mesh fineness in the same time. If we rate the capacity of stamps of 850 pounds at 2 tons per stamp in twenty-four hours, doing the same kind of work, this would show that two sets of 26-inch rolls are equal to 50 stamps. The calculation given in the preceding pages, and results in practice justify, I think, this rating.

Mr. Clark, of the Bertrand Company, says: "We have demonstrated the following facts: That we can pulp more ore with the rollers than can be done with a 40-stamp battery; that we have crushed 9,000 tons of ore without a dollar's expense in repairs; that we have done it with less than one-half the power that would have been required had we used stamps; that the cost of repairs, wear and tear, will not exceed one-quarter of that of crushing with stamps; that we make less dust, jar and noise."

At a later date Mr. Clark says: "We crushed 15,000 tons of ore before putting new tires on the fine-crushing rollers, and have crushed now 20,000 tons; and it will be two or three months yet before we will put new tires on the coarse-crushing rollers. It will be safe to calculate that each set of tires with a set of composition liners for journals will last to crush 20,000 tons of ore. There has been no other expense for repairs upon our rollers except the check-pieces."

On the above basis (crushing 20,000 tons with a wear of two sets of steel tires and one set of composition liners) the cost of renewal at present prices of steel will be as follows:

Two sets of steel tires, ...	$530 00
Freight on 3,264 pounds, at 3 cents, . . .	98 00
Composition liners for journals and check-piece,	100 00
Total	$728 00

which amounts to $3\frac{6}{10}$ cents per ton of ore.

But after all that has been proved for rolls it is still sometimes asserted that they are only suitable for coarse crushing; that ore cannot be pulverized with them sufficiently fine for amalgamation and other purposes where pulverizing is needed, such as cement, phosphate-rock, etc. This is true, as before pointed out, of ordinary cheap rollers only. The same assertion was made when it was proposed to adopt rolls in the Bertrand mill for pulverizing ore for lixiviation; and almost everybody who knew of the affair predicted that the rolls would prove a failure. At the same time, such critics admit that stamp-batteries, centrifugal roller-mills, attrition-mills, ball-pulverizers, mill-stones, and various impracticable machines will pulverize. It would be as good reasoning to argue that a heavy load can be drawn on a sleigh on bare ground, but that the same heavy load cannot be carried on a strong and well-constructed wagon or railroad car. *Crushing with rolls is carrying*

*the load on wheels while all the other devices are similar to
dragging the load with great loss of energy.*

Rolls as a pulverizing machine, if properly constructed, cannot be improved upon either in principle
of operation or in economy of results. All the parts
are of such a character as to contribute to their durability, and there is *no limit* to proportioning the journals
so as to fit them to carry any load that may be put upon
them. The capacity of rolls for quantity and hard work
is far greater than that of any other existing device.

Rolls are at last receiving the recognition they de-
serve as a pulverizing machine. It is now proposed to
employ rolls to crush fine after the ore has passed the
stamp-battery, so as to increase the capacity of an existing plant—which is a complete reversal of all former
ideas on the subject. Yet I have sold rolls for this
purpose. Considering the subject from a mechanical
standpoint it is astonishing that rolls were not brought
to perfection and widely adopted long ago.

KROM'S SYSTEM OF PNEUMATIC CONCENTRATION OF ORES.

Concentration of ores by means of air is a problem which has received much study, and has been the subject of many experiments, both in this country and in Europe. The employment of air for the separation of ores was generally admitted to be a matter of considerable importance, but soon after I had shown that air could be employed successfully as the concentrating agent, and, in fact, had demonstrated its superiority, then some critics claiming to be conversant with the theory and art of ore dressing began to write treatises to prove the impracticability of employing air for this purpose; they said air is "specifically too light, and could move only small particles." These critics evidently had not heard of a cyclone, or the various applications of air as a motive power. In my experience I found that air applied in intermittent impulses, similar to water in the wet-jig, exhibited phenomena favorable to its employment as a concentrating medium. The theory found in literature, and relied upon by the several writers to sustain their criticisms, was that for separating two particles of solids of different density, such liquid medium is the most effective that possesses a density lower than the density of the specifically heavier particles, and higher than the specifically lighter ones. For example (they said), let us suppose that an ore consisting of galena, specific gravity 7.5 and quartz, specific gravity 2.5, so finely crushed that each particle will consist of either galena or gangue to be placed in a liquid of a specific gravity of 5, then it is evident that the quartz would remain floating on the surface, while the galena would sink to the bottom."

A medium of intermediate density is not only impracticable to obtain, but would not serve as the *concentrating medium if we had it.* To separate minerals of different specific gravity, a medium of lower density is required with sufficient motion imparted to it to give the necessary resistance.

The less dense between certain limits that medium is the better it will serve the purpose. In the wet and dry jig it is the resistance caused by the motion of the fluid that effects the separation. In practice we can only use one of two mediums, air or water, all others being chimerical.

A mathematic discussion of the subject would no doubt point out a theoretical medium of such density that its effect taking power, adhesion, etc., into consideration would be a maximum, and that air, instead of water, would be found to be nearest to that theoretical medium. In technical literature it is stated "That the difference between diameters of spheres obtaining equal speed falling in air is less than between diameters of spheres falling equally fast in water." That is, a $\frac{1}{4}''$ globule of galena and $\frac{4}{8}''$ globule of quartz fall with equal speed in still and moving water. But it is asserted that in air the $\frac{4}{8}''$ globule of quartz would fall much faster. This, however, applies to air at rest, which would not comply with the conditions in practice which requires the air to have a motion sufficient to make its resistance equal to water. It is a remarkable circumstance that none of these theoretical writers have comprehended the following facts:

1st. That either air or water can be converted into the exact " theoretical and ideal minimum " by giving motion to it to produce *any degree of resistance.*

2d. That the principle of uniform intermediate resistance does not succeed in practice, but that our most successful practice is based on the intermittent resistance of the concentrating agent.

To prove the incorrectness of the formulas on the subject of the free fall of mineral grains in air and water as applied to ore dressing, I erected two glass tubes (as illustrated) each $2''$ in diameter and $8'$ long. One of these tubes I filled with water and through the other I forced a current of air. I found that practically as stated the $\frac{1}{8}''$ globule of galena and $\frac{3}{8}''$ globule of quartz are equal falling in a column of water, but in a current of air adjusted to retard the galena from falling to the same extent as in water, the $\frac{3}{8}''$ globule of quartz was sustained and did not fall, but oscillated in the open end of the tube.

I also employed bodies of other forms with conical heads (as illustrated in the cut). In experimenting with these I regulated the current of air in the air tube to give the same resistance as water would to the falling bodies of equal weight and size, one in the water, and the other in the current of air, both reaching the bottom in equal time. I then let fall Nos. 1 and 3 in water, and when No. 1 reached the bottom, No. 3 was about 10 behind. Next I let fall the same Nos 1 and 3 in the current of air, and No. 1 fell in the same time in the air current as it did in water, but No. 3 did not fall, but, like the quartz ball, it oscillated in the open end of the tube. Philosophical reasoning would have led to no other conclusion than that the galena and quartz globules would fall with equal velocity in the air current, under the conditions here described. The experiment, however, demonstrated, as practice had before, the superiority of air as the concentrating medium. This experiment has been criticised in a paper written by J. C. Bartlett, A. M. Mr. Bartlett claims that the more favorable effect was due to the narrow cross

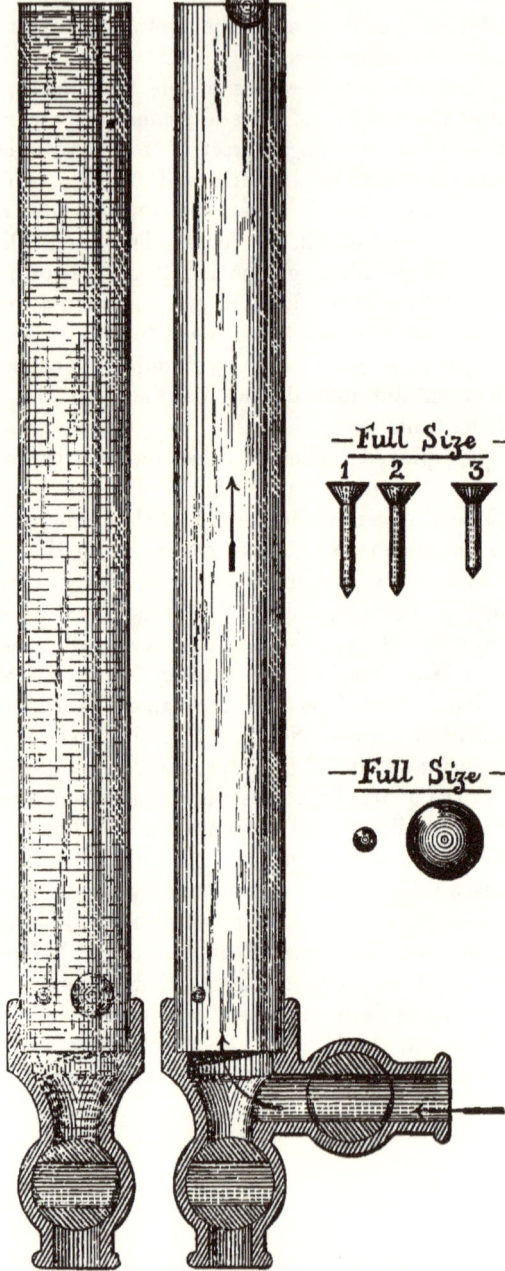

ILLUSTRATING EXPERIMENTS WITH EQUAL FALLING GRAINS
IN WATER AND AIR.

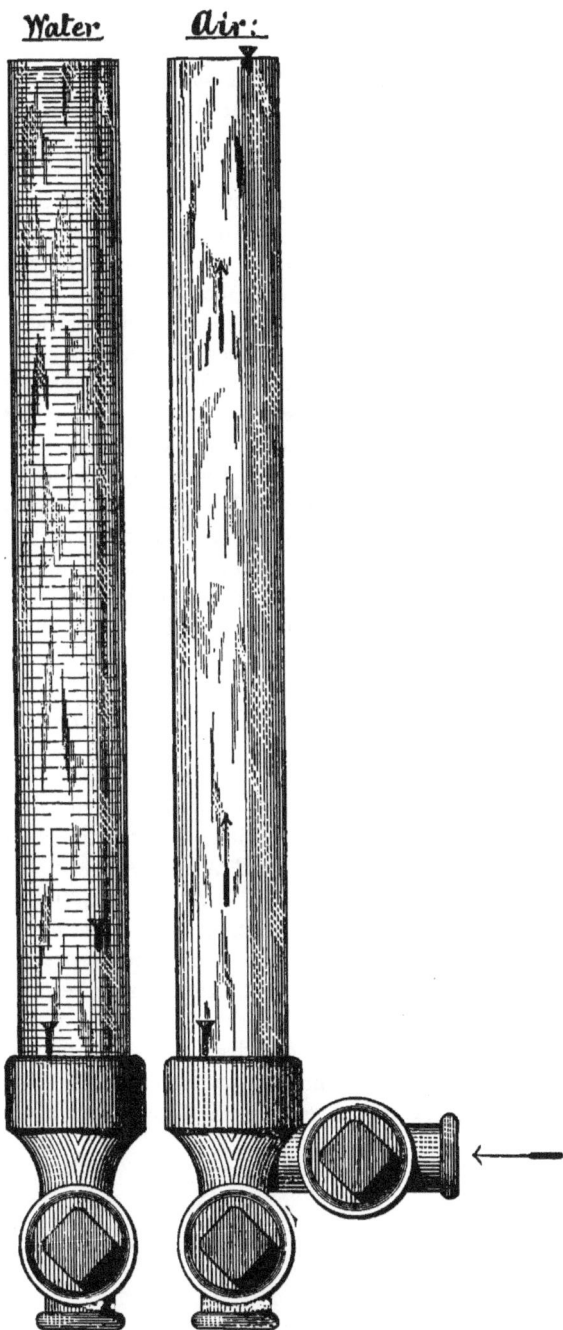

ILLUSTRATING EXPERIMENTS WITH EQUAL FALLING GRAINS
IN WATER AND AIR.

section of the tubes. To avoid such criticisms I performed the second part of the experiment with the conical headed tacks. In this experiment the proper conditions as required by Mr. Bartlett were perfectly complied with, but Mr. Bartlett, however, ignored this part of the experiment. On these criticisms the distinguished Councilor Althaus, one of the Centennial judges, says: "It must not be overlooked that in pneumatic jigging the grains suspended in the stream of air are always in narrow air spaces, and that in cases of the large grains in comparison with the adjacent smaller, but specifically heavy ones, they are similarly situated to those in Krom's experiment in narrow glass tubes. In the opinion of the author, therefore, practical experience alone can decide which process, the hydraulic or pneumatic, can effect the closer separation according to specific gravity and size of grain." These remarks of the councilor are a complete answer to every criticism made against the pneumatic system of concentration.

ENDORSEMENTS OF PNEUMATIC CONCENTRATION.

Dr. R. W. Raymond in the Engineer and Mining Journal, Dec. 21, 1878, in reviewing a treatise on the subject of ore dressing by E. F. Althaus, member of the Royal Mining Court, Breslau, Germany, and judge of group 1, Centennial Exhibition, says: "Councilor Althaus, fortunately for the profession, represented group 1 at the Centennial Exhibition, a member of the jury of that group, and by another piece of good luck or judgment the subject of ore dressing was assigned to him. The treatise on that subject which he contributes to the Reports and Awards is a model of intelligent

criticism, and even as a manual of art superior to anything we possessed in the English language before its
appearance."

The following is taken from Councilor Althaus'
Centennial report on ore dressing:

"THE PRINCIPLES OF KROM'S PNEUMATIC JIGGING."*

"The idea of using air as the medium for the separation of mineral grains, rejected in Europe as fruit-
less, was again taken up in an entirely new direction in
the United States by Stephen R. Krom of New York,
in the year 1868, and has since been systematically
developed in exceedingly ingenious machines, which
were an ornament to the Centennial Exhibition. This
new treatment, only within a few years successfully
introduced into some ore dressing establishments, is yet
too little known in its economical results to give a final
judgment on its possible advantage over ore dressing in
the wet way, but it can even now be designated as a
surprising advance in the art of ore dressing. That the
spirit of invention even in this department has so decidedly succeeded in producing unheard of mechanical effects,
in contradiction to prevailing theoretical opinions, and
furnishing an ocular demonstration so clear that the
correctness of the facts cannot be disputed.

The earlier experiments of air separation were confined to a continuous stream of air produced by a ven-
tilator, in a high, horizontal and narrow conductor (two or
three metres square) through which the dried mineral
grains introduced from above were separated in a manner

* The author had an opportunity at the Centennial Exhibition of observing a Krom machine
at work.

entirely similar to that of the horizontal stream of
water, working in the settling pit according to the
principle of free fall. But Krom starts with the funda-
mental idea of compressing the air by means of a kind
of piston with quick successive strokes, and thus compressed to let it work with great rapidity through a low
stratum of mineral grains, previously assorted according
to size (sized), which rests on fine sieves enclosed laterally in a narrow space. Krom's pneumatic jig or dry
concentrator is similar to the continuous hydraulic jig in
outward appearance, and must be designated as an
intermittent working air stream apparatus, in which,
therefore, the dynamic effect of the living force produced
by an upward jerk through the dense medium has a
more energetic effect than by free fall. In the employment of air in puffs as a medium, its inferior density is
a decided advantage. With water the *vis inertia* of the ·
mass of the medium to be moved prevents driving a
jigger at more than 60 to 120 strokes per minute.
Troublesome setting back of the water takes place,
which acts detrimental to the separation on the sieve,
to avoid which either the stroke and velocity must be
diminished, or a complicated arrangement be made use
of. On the contrary, the air escapes at each stroke
without set back, and it is therefore possible to drive
the pneumatic jigger 420 to 500 strokes per minute,
consequently very much more rapidly than the hydraulic
one.

Air is a medium to be obtained everywhere, while the
procuring of the necessary water supply is attended with
difficulty as a rule, and in many places is impossible.
While the grains to be separated have to pass over a
distance of one or two metres on the continuous
hydraulic jigger, they have to pass over only thirteen
centimetres of sieve length on Krom's machine. This

apparatus can, therefore, for the sake of concentrated efficiency, be more extended laterally without assuming troublesome proportions. As the adhesive effect of air compared with water is extremely small, it is evident that sands can be treated pneumatically which are very much finer than the finest treated on the hydraulic jigger. The dry sand forms a loose mass easily penetrated by the compressed air, while on the hydraulic jigger it is raised altogether in a closely adhering mass by the water, thrust, and even in falling hinders its own separation through mutual adhesion. While grains from .5 to 1 millimetre can scarcely be prepared on the hydraulic jigger, it is said that grains of .01 millimetre can be separated pneumatically."

Prof. Alex. Trippel, at the close of an article on ore dressing published in the *Mining Record*, June 25, '83, says: "We will probably never have dressing works which will save all that is valuable, and work without any disadvantage, but taking all points into consideration, I think that the pneumatic principle as applied by Mr. Krom will gradually be more appreciated as one which is the least wasteful."

T. G. Negus, Esq., the former superintendent of the Clear Creek Co , Col., says:

"There is a very great difference in the quality of ore for concentrating. Some ores are susceptible of very close concentration, while others are not: for instance, some ores contain precious metals in the gangue in such fine particles that disintegration by crushing is impossible. In such ores a high percentage cannot always be obtained. I am prepared to say, however, that the Krom system of dry concentration is eminently superior to all others of which I have any knowledge for all grades of ore."

Mr. Stetefeldt in his recommendations of May 4th,

1883, to the Alta Montana Co., says: "In examining the advantages of the Krom system of dry concentration compared with the wet method, we find, 1st. That Krom's air jig effects a more complete separation of minerals of different specific gravities than the water jig.

2d. That material of greater fineness can be treated in the air jig than in the water jig.

3d. That in wet concentration the great losses occur in those machines which treat material too fine for the jig.

4th. That the dust which results in the Krom system is of higher value than the original ore, and is a concentrated product. But besides this other conditions are to be considered. If the products of concentration are to be roasted and a portion of them have previously to be finer pulverized, it becomes necessary to get them perfectly dry. Now it is evident that it must be very much cheaper, and require a much simpler, and less bulky plant to dry the ore after it leaves the crusher than to remove the moisture from concentrations and slimes completely saturated with water. It furnishes dry dust ready for the roasting furnace, concentrations ready for finer pulverization, a concentrated product ready for smelting, purer and of higher percentage in lead than the water jigs. It also makes possible a better arrangement of the location of the dry kiln in connection with the mill."

Pneumatic concentration has not so far had an opportunity for a long and fair trial; when it does you will see as convincing a demonstration favorable to air as the concentrating medium, as has been shown in favor of employing rollers in place of stamps. By this I do not mean to say that pneumatic concentration has not made a good record, on the contrary, it has shown its superiority whenever and wherever tested. In the Clear

Creek Company's Mill at Georgetown, Colorado, the
pneumatic jigs gave 10 to 15 per cent. better results
than tests on the same ore in competition with wet concentration. This mill was kept in operation as long as
ore could be obtained. At Galena, Nevada, in the
White & Shiloh Mining Company's Mill it gave results
satisfactory to the owners, and above the guaranteed
percentage of saving, and on such ore which previous
trials had shown could not be successfully separated by
the wet process, and this works remained in successful
operation until the mine produced'no more ore.

At Austin, Nevada, the results were decidedly in
favor of pneumatic concentration. At Wickes, Montana,
the dry concentration works were destroyed by fire
before they had been put in proper running order, or
placed in charge of persons competent to run them.

Also the works at Star Canyon, Nevada. The first
mill erected on the pneumatic system have always
been considered a success. It is purely prejudice and
the misinterpretation of the theories applicable to ore
dressing and accidental circumstances which has prevented
the more rapid introduction of the pneumatic system.

THE PNEUMATIC JIG.

The first plate, Fig. 1, is a perspective view of the pneumatic jig. Fig. 2 is an end view, and Fig. 3 a transverse sectional view. The machine is composed essentially of the following parts: A receiver, II, to hold the crushed ore; and ore bed, O, on which the ore is submitted to the actions of the air; the two gates, G, G, one to regulate the flow of ore from the receiver II, the other to determine the depth of ore on the ore bed; passage C, in which the concentrated ore descends, and roller R, to effect and regulate the discharge of the same; a fan, B, to give the puffs of air, a trip-wheel T, lever L, and spring S, to operate the fan, and a ratchet-wheel W, and pawl P, to impart revolution to the roller R.

The mode of operating the machine is as follows: Ore is placed in the receiver II, and the driving pulley set in motion. The cam shaped trip-wheel T, fixed on the opposite end of the pulley-shafts, works against the lever L. By the alternate action of this wheel, forcing the lever in one direction, and the spring which suddenly carries it back again, the fan B is made to swing on the shaft I, sending at each upward movement a quick and sharp puff of air through the ore bed, and lifting slightly the ore lying on it. There are six projections upon the trip-wheel, so that the moderate speed of 80 to 90 revolutions per minute will give 480 to 540 upward movements of the fan in the same time, and a corresponding number of puffs of air to agitate the ore. This rate is sufficient to secure a steady motion of the heavy pulley, and yet not so fast as to produce any perceptible vibration, the machine working smoothly and easily. The ore bed is composed of wire-gauze tubes, placed at distances from each other of $\frac{3}{16}$, $\frac{1}{4}$, $\frac{3}{8}$ and $\frac{1}{2}$ of an inch, according to the grade of ore to be concentrated, the finer

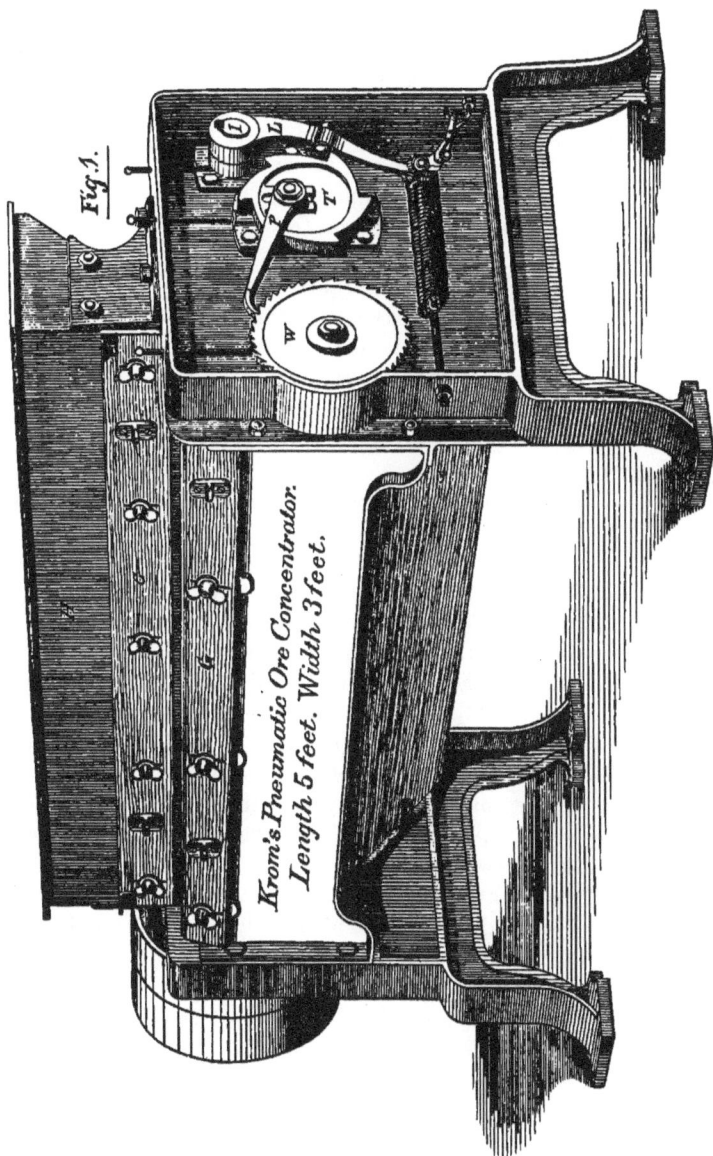

Fig. 1.

Krom's Pneumatic Ore Concentrator.
Length 5 feet. Width 3 feet.

Patented Aug. 4, 1868, and Sept. 1, 1868; Re-Issue Nov. 3, 1868; Re-Issue June 20, 1871; Patented
Dec. 5, 1871, and Patents pending for improvements.

requiring the tubes set nearer together, while the coarser allow the tubes to be placed farther apart. The ore bed (situated in front of the fan, as plainly shown in the sectional view) is composed of these tubes. Their ends next to the fan being open, the air from the bellows enters and escapes through the top and sides of the tubes, agitating the ore that lies on them, and also that between them near the surface.

The ore between the tubes rests on that immediately underneath in the passage C, and sinks as fast as the roller R effects its discharge. The tubes being open also on the bottom, any fine ore passing through the meshes of the wire gauze descends with the main body C, thus preventing any liability of the tubes to fill up with fine ore.

The roller R is operated (as above mentioned) by means of the ratchet-wheel W and pawl P, and the latter being carried by a crank on the trip-wheel, it follows that its speed is governed by the speed of this wheel, which also gives motion to the fan B. By this connection the fan, which effects the concentration, and the roller, which discharges the concentrated ore, are made to act in concert with each other. The importance of this feature will be apparent when it is remembered that the amount of ore concentrated in a given time depends on the rapidity of the puffs of air, so that the motion of the discharge roller R should be regulated to correspond with the speed of the fan.

The crank which carries the pawl can also be varied in length so that the speed of the rollers may be regulated according to the richness of the ore. As already stated, the upper gate, G, governs the flow of ore from the receiver, H, to the ore bed, while the lower gate, G, regulates the thickness of the stratum of ore lying on the latter, as it is necessary to increase or diminish the depth of the

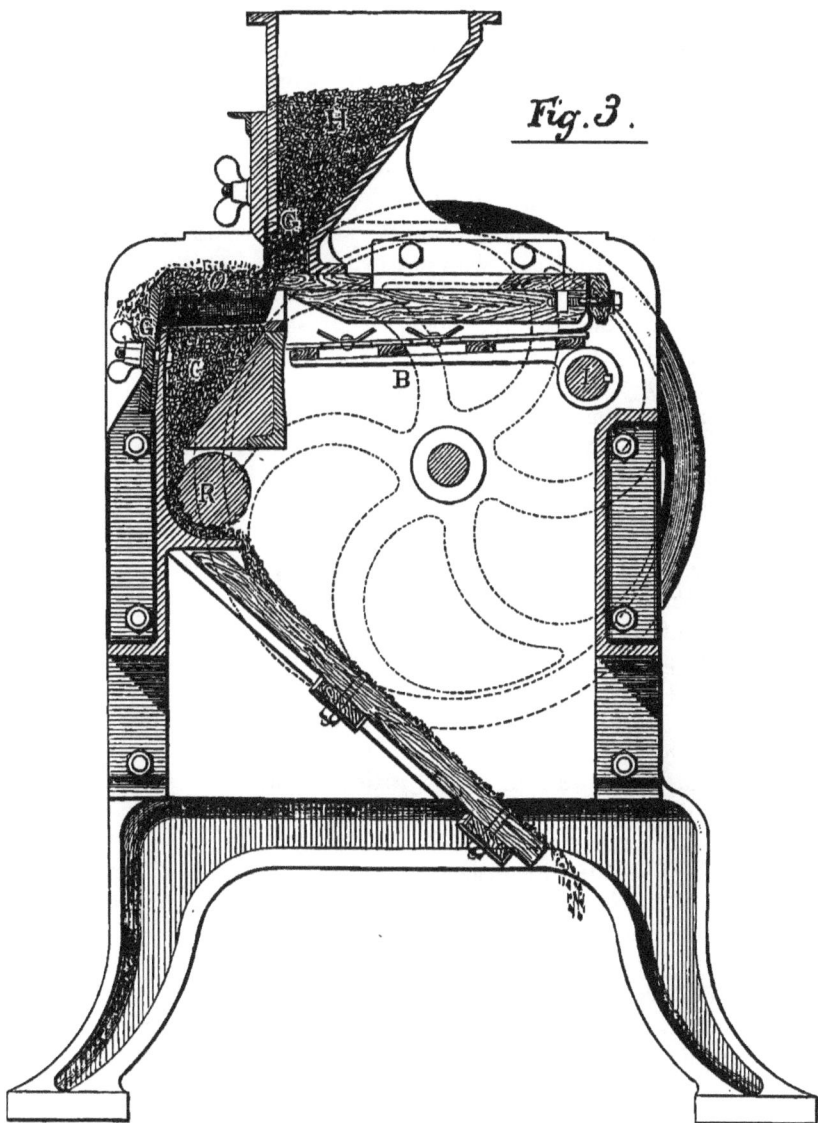

Fig.3.

SECTIONAL VIEW OF KROM'S PNEUMATIC JIG.

bed of ore operated upon according to its coarseness or fineness. The finer the crushed ore, the thinner the stratum must be. The strap with its screw fastenings, serves to prevent the roller attachment of the lever L from striking the body of the trip-wheel, as it falls from each of the cam-shaped projections, and to regulate the extent of movement of the fan. That is, the strap must in all cases be so adjusted that the small roller working against the trip-wheel shall not strike at the foot of the cam—the strap serving in this manner to cushion the blow. Further, by tightening up or slacking off, by means of the screw fastening, the fan is carried in its vibration through a greater or less space, producing a stronger or lighter puff of air. It will be understood that the volume of the puff of air required varies with each grade of ore operated upon. Now, with the strap arrangement alone, the puff of air can be regulated to the requirements of different grades of ore. But as the finer grades demand so much less movement of the fan than do the coarser, it is better to select a trip-wheel from the sizes furnished which gives a movement corresponding most nearly to that required, and then to make the closer adjustments by means of the screw and strap; *but the roller must in no instance strike at the foot of the cam.*

The novel features to be particularly noted in this separator, are as follows :

1st. The ore-bed.

2d. The automatic discharge rollers.

3d. The fan for producing the puffs of air.

4th. The trip-wheel and springs.

5th. The strap and adjustable screw.

6th. the gate on the receiver H.

1. The ore-bed, formed of wire gauze tubes which are set in a frame at short distances apart, to allow the ore

to sink between them, is a novel device for securing the removal of the concentrated ore as fast as the separation on the bed is completed.

The entire width of the ore-bed is 4 feet, and along this tubes only $\frac{1}{2}$ inch wide are set with intervening spaces of $\frac{3}{16}$ to $\frac{1}{2}$ inch, consequently the total extent of the spaces through which the ore sinks is $\frac{1}{3}$ to $\frac{1}{2}$ of the entire ore-bed. By this arrangement the downward flow is very gradual between the tubes.

The automatic discharge roller R—This being driven by the same motion that works the fan, it follows that the concentration of the ore and its discharge are effected in concert, so that when the speed of the machine slackens, the concentration being less, the rate of delivery is correspondingly reduced.

The device for producing the puffs of air, namely—A fan directly actuated by means of a single lever, L, is very simple and effective.

The trip-wheel and springs actuatings the fan or bellows—The wheel by the gradual action of its cams throws back the lever L, and consequently the fan or bellows-plate downward with a movement as gradual as possible, and immediately the springs carry the fan quickly upward.

The superiority of this device over cams, cranks, etc., is, that with it a considerable variation of the speed of the machine does not effect the quality of the concentration, but only the quantity.

If the trip-wheel revolves slowly the number of vibrations of the fan is less, but as the spring causes the upward movement of the fan the puff of air acts with uniform force, and we thereby obtain at all speeds what we may term a concentrating stroke of the bellows, as the more sharply the puffs of air are given the more perfect will be the separation.

In reference to the strap and adjustable screw, the explanation already given is sufficient.

The gate on the hopper, H, compels the ore to flow on the ore-bed as an under-current, and as the puffs are regulated to agitate very slightly the heavy particles, only the lighter portions will rise to the surface and be thrown over the lower gate, G, as tailings, while the heavy will sink through the ore-bed to be discharged by the roller R.

'This feature enables us to concentrate perfectly with a very short travel of the ore, or in other words, to employ a short ore-bed which is of great advantage.

By having a short ore-bed we can extend what we properly term the width of the bed, thereby greatly increasing the capacity of a machine of a given size. All other experimenters have caused the discharge of the ore to take place over the narrow side of the machine and the travel of the ore over the greater distance. The reverse of this takes place in this machine, viz.: the distance of travel of the ore over the bed is only 5 inches, while the line of overflow is 4 feet, and can be, at pleasure, made still greater. A short bed enables us to use a small fan, and reduces the vibrations attending rapid movements. A more even and uniform agitation is secured when the ore is confined within narrow limits, and more satisfactory results are obtained.

Since I have discovered that a short ore-bed, of only 5 to 6 inches in length, is not only sufficient, but, in fact, much superior, and that the width of the bed and extent of overflow can accordingly be increased, I am able so to place the fan, and to group in compact form all the working parts, and to very considerably reduce the size of the machine.

The puffs of air are regulated to agitate sufficiently the ore on the bed, but should the richness of the ore

increase during working and too large an amount collect
on the bed, the air ceases to lift or agitate the material
as much, and so a check is furnished to prevent loss in
the tailings. No such check is possible in water concentration, because water moves practically as a solid and
carries all before it.

All parts in this machine liable to wear are manufac-
tured in duplicate and can be cheaply replaced. The
machine measures 6 feet in length over all, 3 feet in
width, and 3 feet 10 inches in height. Its weight, complete, is 1,200 pounds, and it is capable of concentrating
$\frac{1}{2}$ ton per hour with $\frac{1}{8}$ horse power.

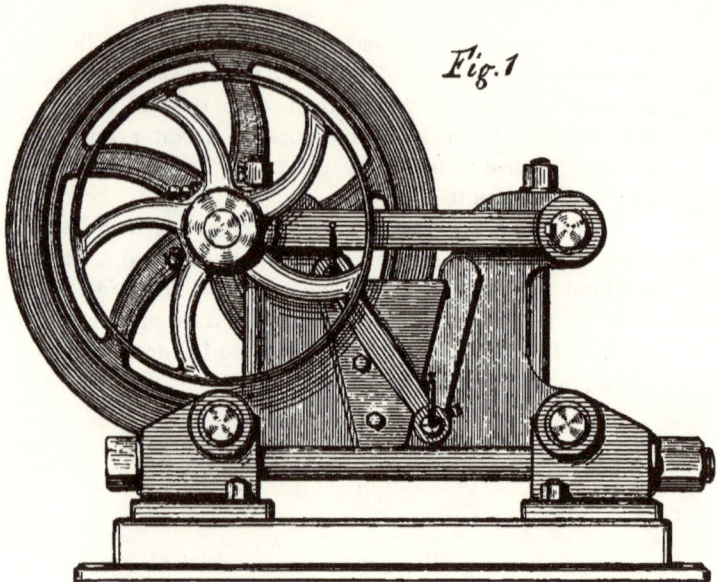

Fig. 1

KROM'S PATENTED LABORATORY CRUSHER.

The foregoing cut represents a crusher for laboratory use.

In this machine both jaws oscillate on centers fixed some distance from the crushing faces. The principal feature is the employment of segments of circles between which the ore is crushed on the same principle as rollers act.

The lower ends of the crushing plates are true segments of circles, and throughout all the movements of the jaws they remain at fixed distances from each other, but the top parts of the plates recede from each other with straight lines. The crusher can be adjusted, by means of bolts, so as to produce either fine or coarse material. The crushing faces are made of steel, and the lower half of the steel plates are hardened. It is the only machine in the market adapted for laboratory use.

LABORATORY JAW CRUSHER.

LABORATORY ROLLERS.

Fig. 2.

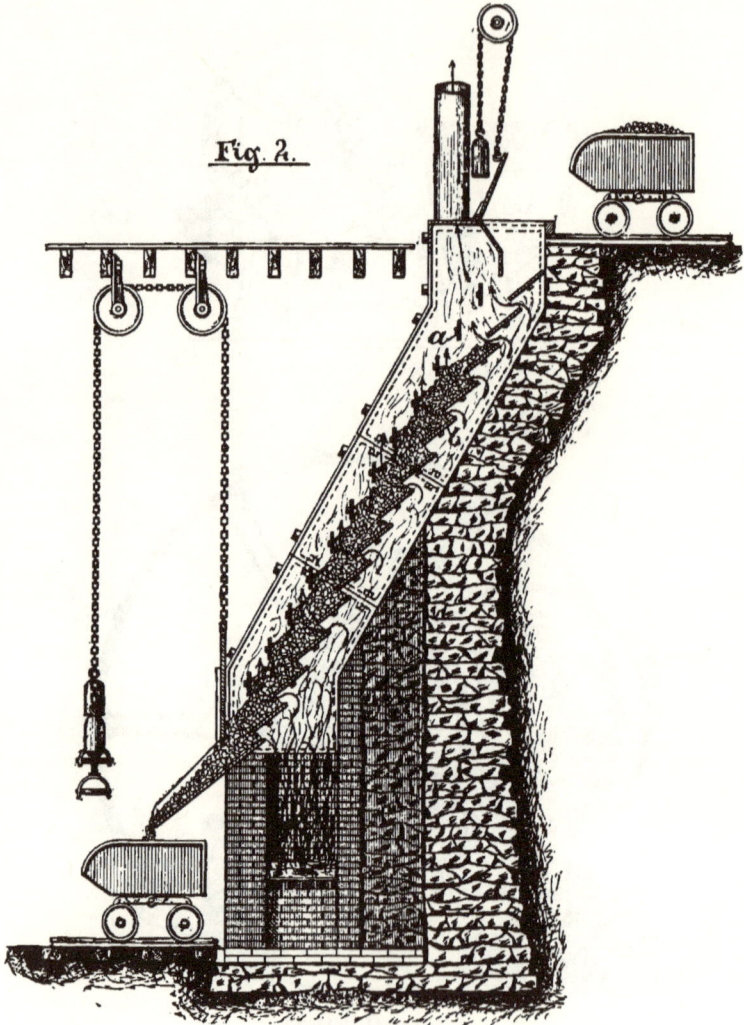

KROM'S DRY KILN.

Patent for this Dry Kiln pending.

KROM'S DRY KILN.

The cut, Fig. 2, illustrates my improved dry kiln for drying ores after it is broken by an ore breaker. The cast-iron plates, *b*, on which the ore rests while drying, are arranged in steps, with spaces between each step of 3 or 4 inches. These spaces allow the hot air and gases from the fire underneath to pass up through the strata of coarsely crushed ore, as plainly indicated by the arrows.

The waste heat and evaporated moisture pass out through the chimney. The plates, *b*, or steps, are placed at an angle of 45°, but it will be observed that the furnace assumes an angle of 58°. Therefore to maintain the strata of ore of a uniform thickness it is necessary to have the check plates, *a*.

The distance between the lower edge of these plates, *a*, and the plates *b*, determines the thickness of ore strata. This thickness can range from 6 to 9 inches. The check plates, *a*, can be readily varied in height by means of holes in the flanges which support them.

If the location of the mill site will admit of so much elevation, the plan shown in Fig. 1 will save much labor and expense in handling the ore. It will be seen that the plan here suggested is for continuous and automatic operation.

Feed roller *c* is to control and regulate the flow of ore to the crushing rollers. It has become necessary to employ some means for collecting the pieces of iron and steel before they pass to the pulverizing machines.

A system of magnets is placed in the shute *a, z, d*.

The arrangement of the magnets are more plainly shown in Figs. 4 and 5.

This dry kiln has decided advantages over the revolving drier. The kiln here illustrated requires no power to operate it, and the gentle flow of ore by gravity over the plates does not tend to stir up dust or create more from abrasion, and consequently no dust chamber is required as with cylinder drier.

The fuel required to dry ore in this kiln is the minimum amount. It is 20 feet long, 5 feet wide, and holds a strata of ore 6 to 8 inches thick. The capacity of this kiln may be estimated between 2½ to 5 tons per hour.

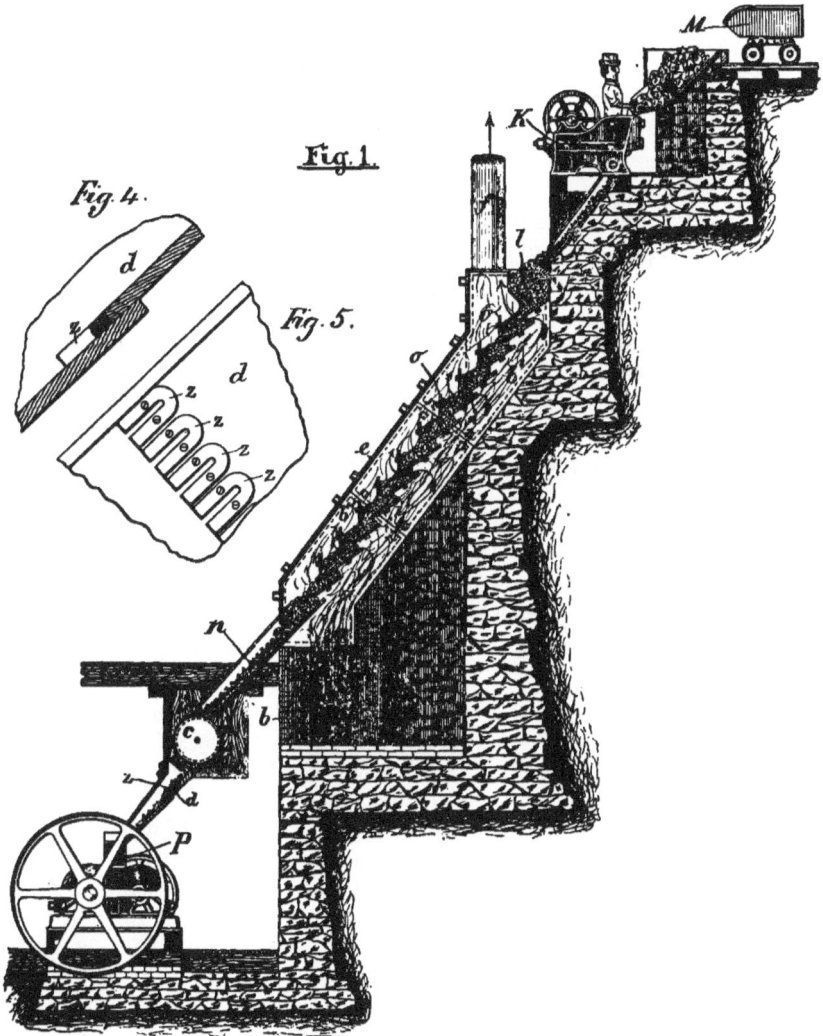

Fig. 4.

Fig. 1.

Fig. 5.

Patent for this Dry Kiln pending.

KROM'S REVOLVING SCREEN.

The cut on the following page represents my improved screen frame. The frame is composed of light wrought iron angle-bars, rolled expressly for the manufacture of this screen. The form of this bar is shown in the sectional cut of the rail marked *k*. The spiders which carry these angle-bars are composed of a hub and wrought iron spokes. The spokes are screwed into the hub to a shoulder formed on the spoke. On the outer end of each spoke is another shoulder on which the angle-bars rest. All the parts are made of uniform dimensions, and are therefore interchangeable. The angle-bars hold the screen-head in the small end by means of small bolts.

The iron framework of the screen, with its clamps, are independent of the wooden frame on which the wire cloth is nailed. To fasten the screen-frame in place it is simply slid under the clamps, and screwed down to the angle-bars.

The clamps which hold the screen-frames to the angle-bars are all held up with springs, so that when they are released the clamps are carried clear from the screen-frame, and the same is taken out and put in without difficulty by one person.

The iron angle-bars carry bolts on which are sliding weights to jar the screen. On the bolt, next to the screen, I put a rubber cushion, so that a sharp blow is given when the weight falls from the top of the screen, and a soft blow when the weight falls on the lower side of the screen. The object of the jar is to keep the meshes of the wire-cloth free from particles of ore. I also sometimes provide a casing, *o*, for the sliding weight, to protect fine screens from damage, which may result from the breaking of the bolts on which the weight

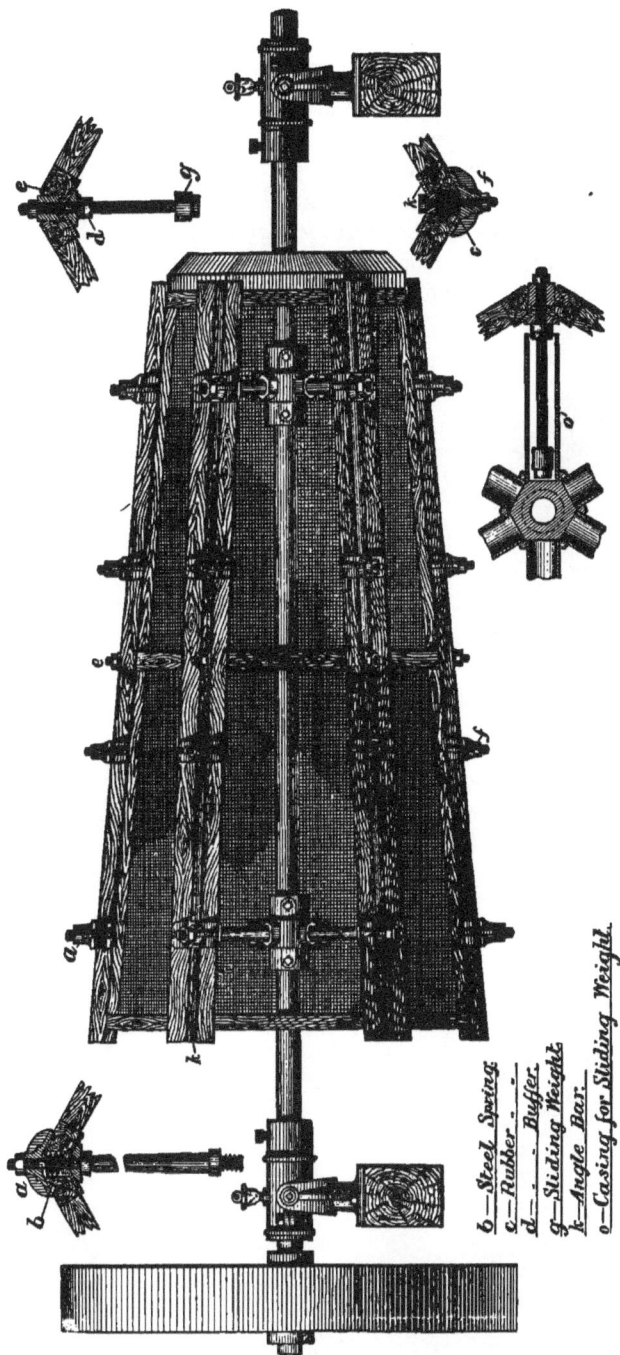

b—*Steel Spring.*
c—*Rubber - - .*
d—- , - *Buffer.*
g—*Sliding Weight.*
k—*Angle Bar.*
o—*Casing for Sliding Weight.*

KROM'S REVOLVING SCREEN.
Patent for this Screen allowed September 11th, 1885.

slides. These casings are made of thin iron tubes screwed in a hub, similar to the hubs in which the spokes of the screen are secured. The tubes are large enough to clear the sliding or falling weight, and long enough to enclose the bolt nearly its whole length. In case the bolt breaks the tubes act as a pocket to hold it. This screen is more complete and perfect in all its arrangements and details than any other in the market.

www.ingramcontent.com/pod-product-compliance
Lightning Source LLC
Chambersburg PA
CBHW022028080426
42733CB00007B/766